# A Fortnight of Daily Mathematics

Suggested usage: Keep the book blank, and just copy the pages that you want to use.

\* \* \*

(blank page)

# Mathematics Exercise for Day 1

Name: _____ Date: _____

1. $5/(1/2 + 1/3)$

2. $-3 - -10$

3. $x^5 - x^3 + 3x + y + 1 = x^5 + y + 16 - x^3$

4. $x^3 + abc + (x - 51)(x - 87) = abc + x^3$

5. $x^{1000} + x^2 = 47x + x^{1000}$

(end of document)

(blank page)

# Mathematics Exercise for Day 2

Name: _____     Date: _____

1.  $7/(1/3 + 1/4)$

2.  $-4 - -15$

3.  $x^5 - x^3 + 3x + y + 1 = x^5 + y + 25 - x^3$

4.  $x^3 + abc + (x - 42)(x - 61) = abc + x^3$

5.  $x^{1000} + x^2 = 35x + x^{1000}$

(end of document)

(blank page)

# Mathematics Exercise for Day 3

Name: _____     Date: _____

1.    $9/(1/4 + 1/5)$

2.    $-8 - -20$

3.    $x^5 - x^3 + 5x + y + 1 = x^5 + y + 11 - x^3$

4.    $x^3 + abc + (x - 15)(x - 27) = abc + x^3$

5.    $x^{1000} + x^2 = 94x + x^{1000}$

(end of document)

(blank page)

# Mathematics Exercise for Day 4

Name: _____    Date: _____

1.  $11/(1/5 + 1/6)$

2.  $-1 - -30$

3.  $x^5 - x^3 + 5x + y + 1 = x^5 + y + 36 - x^3$

4.  $x^3 + abc + (x - 11)(x - 83) = abc + x^3$

5.  $x^{1000} + x^2 = 89x + x^{1000}$

(end of document)

(blank page)

# Mathematics Exercise for Day 5

Name: _____    Date: _____

1.  $13/(1/6 + 1/7)$

2.  $-2 - -25$

3.  $x^5 - x^3 + 6x + y + 1 = x^5 + y + 7 - x^3$

4.  $x^3 + abc + (x - 21)(x - 28) = abc + x^3$

5.  $x^{1000} + x^2 = 15x + x^{1000}$

(end of document)

(blank page)

# Mathematics Exercise for Day 6

Name: _____     Date: _____

1.  $15/(1/7 + 1/8)$

2.  $-8 - -40$

3.  $x^5 - x^3 + 2x + y + 1 = x^5 + y + 21 - x^3$

4.  $x^3 + abc + (x - 48)(x - 49) = abc + x^3$

5.  $x^{1000} + x^2 = 99x + x^{1000}$

(end of document)

(blank page)

# Mathematics Exercise for Day 7

Name: _____     Date: _____

1. $17/(1/8 + 1/9)$

2. $-3 - -40$

3. $x^5 - x^3 + 8x + y + 1 = x^5 + y + 17 - x^3$

4. $x^3 + abc + (x - 30)(x - 41) = abc + x^3$

5. $x^{1000} + x^2 = 62x + x^{1000}$

(end of document)

(blank page)

# Mathematics Exercise for Day 8

Name: _____          Date: _____

1.  3/(1/2 + 1/4)

2.  -7 - -50

3.  $x^5 - x^3 + 2x + y + 1 = x^5 + y + 31 - x^3$

4.  $x^3 + abc + (x - 18)(x - 55) = abc + x^3$

5.  $x^{1000} + x^2 = 73x + x^{1000}$

(end of document)

(blank page)

# Mathematics Exercise for Day 9

Name: _____     Date: _____

1. $7/(1/2 + 1/5)$

2. $-4 - -50$

3. $x^5 - x^3 + 4x + y + 1 = x^5 + y + 21 - x^3$

4. $x^3 + abc + (x - 25)(x - 76) = abc + x^3$

5. $x^{1000} + x^2 = 13x + x^{1000}$

(end of document)

(blank page)

# Mathematics Exercise for Day 10

Name: _____     Date: _____

1.  $4/(1/2 + 1/6)$

2.  $-5 - -30$

3.  $x^5 - x^3 + 9x + y + 1 = x^5 + y + 19 - x^3$

4.  $x^3 + abc + (x - 12)(x - 14) = abc + x^3$

5.  $x^{1000} + x^2 = 80x + x^{1000}$

(end of document)

(blank page)

# Mathematics Exercise for Day 11

Name: _____     Date: _____

1.  9/(1/2 + 1/7)

2.  -9 - -30

3.  $x^5 - x^3 + 2x + y + 1 = x^5 + y + 17 - x^3$

4.  $x^3 + abc + (x - 31)(x - 42) = abc + x^3$

5.  $x^{1000} + x^2 = 81x + x^{1000}$

(end of document)

(blank page)

# Mathematics Exercise for Day 12

Name: _____     Date: _____

1. $5/(1/2 + 1/8)$

2. $-1 - -50$

3. $x^5 - x^3 + 8x + y + 1 = x^5 + y + 65 - x^3$

4. $x^3 + abc + (x - 77)(x - 16) = abc + x^3$

5. $x^{1000} + x^2 = 32x + x^{1000}$

(end of document)

(blank page)

# Mathematics Exercise for Day 13

Name: _____     Date: _____

1.  $11/(1/2 + 1/9)$

2.  $-1 - -32$

3.  $x^5 - x^3 + 5x + y + 1 = x^5 + y + 41 - x^3$

4.  $x^3 + abc + (x - 20)(x - 59) = abc + x^3$

5.  $x^{1000} + x^2 = 29x + x^{1000}$

(end of document)

(blank page)

# Mathematics Exercise for Day 14

Name: _____     Date: _____

1. $8/(1/3 + 1/5)$

2. $-3 - -35$

3. $x^5 - x^3 + 4x + y + 1 = x^5 + y + 45 - x^3$

4. $x^3 + abc + (x - 19)(x - 52) = abc + x^3$

5. $x^{1000} + x^2 = 91x + x^{1000}$

(end of document)

Made in the USA
Lexington, KY
28 March 2018